Nuttall's Table Book

REVISED AND ENLARGED BY
HELEN PERRIN, B.Sc., M.Sc.

INCLUDING NEW SECTIONS ON
DECIMAL CURRENCY AND METRIC UNITS

Main Sections

ARITHMETIC AND NUMBER
CURRENCY AND COMMERCE
WEIGHTS AND MEASURES
TIME AND CALENDAR
GENERAL INFORMATION

FREDERICK WARNE & CO LTD
London · New York

Contents—Index

Printed in Great Britain by William Clowes & Sons, Limited
London, Beccles and Colchester 755.174

ARITHMETIC AND NUMBER

Arithmetic

Arithmetic is the science of numbers and the art of computing by numbers. The name is derived from the Greek word 'arithmos', meaning number. It may be considered as the most useful of all the arts and was used by the remotest ages of civilisation. All calculation in arithmetic is done with the aid of one or more of the four following rules or processes.

1. Addition (+) by which several numbers or sums are added together to make a whole, or total, sum. When numbers which are to be added consist of units of several degrees, it is found convenient to add the units of each degree by themselves; and since ten units of any degree make one unit of the next higher degree, the number of tens in the sum of each degree of units is carried to the next higher degree and added to it. So to add several numbers they are written one under the other with the units under units, tens under tens, and so on.

e.g. 217 + 50498 + 2556 is written

```
  217
50498
 2556
-----
53271
```

and the total is 53271

2. Subtraction (−) by which a lesser number is taken from a greater. What is left is called the **difference** of the two numbers or **remainder**.

3. Multiplication (×) which ascertains what a number will amount to when added a certain number of times.

e.g. $13 \times 6 =$ 13 added 6 times
$= 13 + 13 + 13 + 13 + 13 + 13 = 78$

The **multiplicand** is the number to be multiplied, the **multiplier** is the number that multiplies and the **product** is the result of the multiplication. The product of small numbers may be committed to memory; but when the product of factors consisting of several figures is required, it is necessary to multiply each separate figure in the multiplicand by each figure in the multiplier. Care must be taken to denote the several products in such order that they represent their respective values, placing the tens under tens, hundreds under hundreds, etc., so that they can be added up.

e.g. 2474321 × 125 is worked out:

	2474321
	125
Units......	12371605
Tens......	4948642
Hundreds..	2474321
The result is.....	309290125

4. Division (÷) which ascertains how many times a lesser number is contained in a greater. The **dividend** is the number to be divided, the **divisor** is the number by which the division is done and the **quotient** shows how many times the divisor is contained in the dividend. The remainder is the number left over; if there is one, it is always less than the divisor.

e.g. 45÷7: 45 − 7 − 7 − 7 − 7 − 7 − 7 = 3,

7 can be subtracted 6 times, leaving 3 over, so 45÷7 = 6 remainder 3.

Addition and Subtraction Table

1 and	2 and	3 and	4 and	5 and	6 and
1 are 2	1 are 3	1 are 4	1 are 5	1 are 6	1 are 7
2 ,, 3	2 ,, 4	2 ,, 5	2 ,, 6	2 ,, 7	2 ,, 8
3 ,, 4	3 ,, 5	3 ,, 6	3 ,, 7	3 ,, 8	3 ,, 9
4 ,, 5	4 ,, 6	4 ,, 7	4 ,, 8	4 ,, 9	4 ,, 10
5 ,, 6	5 ,, 7	5 ,, 8	5 ,, 9	5 ,, 10	5 ,, 11
6 ,, 7	6 ,, 8	6 ,, 9	6 ,, 10	6 ,, 11	6 ,, 12
7 ,, 8	7 ,, 9	7 ,, 10	7 ,, 11	7 ,, 12	7 ,, 13
8 ,, 9	8 ,, 10	8 ,, 11	8 ,, 12	8 ,, 13	8 ,, 14
9 ,, 10	9 ,, 11	9 ,, 12	9 ,, 13	9 ,, 14	9 ,, 15
10 ,, 11	10 ,, 12	10 ,, 13	10 ,, 14	10 ,, 15	10 ,, 16
11 ,, 12	11 ,, 13	11 ,, 14	11 ,, 15	11 ,, 16	11 ,, 17
12 ,, 13	12 ,, 14	12 ,, 15	12 ,, 16	12 ,, 17	12 ,, 18
1 from	**2 from**	**3 from**	**4 from**	**5 from**	**6 from**

7 and	8 and	9 and	10 and	11 and	12 and
1 are 8	1 are 9	1 are 10	1 are 11	1 are 12	1 are 13
2 ,, 9	2 ,, 10	2 ,, 11	2 ,, 12	2 ,, 13	2 ,, 14
3 ,, 10	3 ,, 11	3 ,, 12	3 ,, 13	3 ,, 14	3 ,, 15
4 ,, 11	4 ,, 12	4 ,, 13	4 ,, 14	4 ,, 15	4 ,, 16
5 ,, 12	5 ,, 13	5 ,, 14	5 ,, 15	5 ,, 16	5 ,, 17
6 ,, 13	6 ,, 14	6 ,, 15	6 ,, 16	6 ,, 17	6 ,, 18
7 ,, 14	7 ,, 15	7 ,, 16	7 ,, 17	7 ,, 18	7 ,, 19
8 ,, 15	8 ,, 16	8 ,, 17	8 ,, 18	8 ,, 19	8 ,, 20
9 ,, 16	9 ,, 17	9 ,, 18	9 ,, 19	9 ,, 20	9 ,, 21
10 ,, 17	10 ,, 18	10 ,, 19	10 ,, 20	10 ,, 21	10 ,, 22
11 ,, 18	11 ,, 19	11 ,, 20	11 ,, 21	11 ,, 22	11 ,, 23
12 ,, 19	12 ,, 20	12 ,, 21	12 ,, 22	12 ,, 23	12 ,, 24
7 from	**8 from**	**9 from**	**10 from**	**11 from**	**12 from**

Note—This table may be applied to subtraction by reversing it thus: 3 from 7 leaves 4; 9 from 14 leaves 5.

Multiplication Tables

Twice	3 times	4 times	5 times	6 times	7 times	8 times
1 are 2	1 are 3	1 are 4	1 are 5	1 are 6	1 are 7	1 are 8
2 ,, 4	2 ,, 6	2 ,, 8	2 ,, 10	2 ,, 12	2 ,, 14	2 ,, 16
3 ,, 6	3 ,, 9	3 ,, 12	3 ,, 15	3 ,, 18	3 ,, 21	3 ,, 24
4 ,, 8	4 ,, 12	4 ,, 16	4 ,, 20	4 ,, 24	4 ,, 28	4 ,, 32
5 ,, 10	5 ,, 15	5 ,, 20	5 ,, 25	5 ,, 30	5 ,, 35	5 ,, 40
6 ,, 12	6 ,, 18	6 ,, 24	6 ,, 30	6 ,, 36	6 ,, 42	6 ,, 48
7 ,, 14	7 ,, 21	7 ,, 28	7 ,, 35	7 ,, 42	7 ,, 49	7 ,, 56
8 ,, 16	8 ,, 24	8 ,, 32	8 ,, 40	8 ,, 48	8 ,, 56	8 ,, 64
9 ,, 18	9 ,, 27	9 ,, 36	9 ,, 45	9 ,, 54	9 ,, 63	9 ,, 72
10 ,, 20	10 ,, 30	10 ,, 40	10 ,, 50	10 ,, 60	10 ,, 70	10 ,, 80
11 ,, 22	11 ,, 33	11 ,, 44	11 ,, 55	11 ,, 66	11 ,, 77	11 ,, 88
12 ,, 24	12 ,, 36	12 ,, 48	12 ,, 60	12 ,, 72	12 ,, 84	12 ,, 96
13 ,, 26	13 ,, 39	13 ,, 52	13 ,, 65	13 ,, 78	13 ,, 91	13 ,, 104
14 ,, 28	14 ,, 42	14 ,, 56	14 ,, 70	14 ,, 84	14 ,, 98	14 ,, 112
15 ,, 30	15 ,, 45	15 .. 60	15 ,, 75	15 ,, 90	15 ,, 105	15 ,, 120
16 ,, 32	16 ,, 48	16 ,, 64	16 ,, 80	16 ,, 96	16 ,, 112	16 ,, 128
17 ,, 34	17 ,, 51	17 ,, 68	17 ,, 85	17 ,, 102	17 ,, 119	17 ,, 136
18 ,, 36	18 ,, 54	18 ,, 72	18 ,, 90	18 ,, 108	18 ,, 126	18 ,, 144
19 ,, 38	19 ,, 57	19 ,, 76	19 ,, 95	19 ,, 114	19 ,, 133	19 ,, 152
20 ,, 40	20 ,, 60	20 ,, 80	20 ,, 100	20 ,, 120	20 ,, 140	20 ,, 160

Multiplication Tables cont.

9 times	10 times	11 times	12 times	13 times	14 times	15 times
1 are 9	1 are 10	1 are 11	1 are 12	1 are 13	1 are 14	1 are 15
2 ,, 18	2 ,, 20	2 ,, 22	2 ,, 24	2 ,, 26	2 ,, 28	2 ,, 30
3 ,, 27	3 ,, 30	3 ,, 33	3 ,, 36	3 ,, 39	3 ,, 42	3 ,, 45
4 ,, 36	4 ,, 40	4 ,, 44	4 ,, 48	4 ,, 52	4 ,, 56	4 ,, 60
5 ,, 45	5 ,, 50	5 ,, 55	5 ,, 60	5 ,, 65	5 ,, 70	5 ,, 75
6 ,, 54	6 ,, 60	6 ,, 66	6 ,, 72	6 ,, 78	6 ,, 84	6 ,, 90
7 ,, 63	7 ,, 70	7 ,, 77	7 ,, 84	7 ,, 91	7 ,, 98	7 ,, 105
8 ,, 72	8 ,, 80	8 ,, 88	8 ,, 96	8 ,, 104	8 ,, 112	8 ,, 120
9 ,, 81	9 ,, 90	9 ,, 99	9 ,, 108	9 ,, 117	9 ,, 126	9 ,, 135
10 ,, 90	10 ,, 100	10 ,, 110	10 ,, 120	10 ,, 130	10 ,, 140	10 ,, 150
11 ,, 99	11 ,, 110	11 ,, 121	11 ,, 132	11 ,, 143	11 ,, 154	11 ,, 165
12 ,, 108	12 ,, 120	12 ,, 132	12 ,, 144	12 ,, 156	12 ,, 168	12 ,, 180
13 ,, 117	13 ,, 130	13 ,, 143	13 ,, 156	13 ,, 169	13 ,, 182	13 ,, 195
14 ,, 126	14 ,, 140	14 ,, 154	14 ,, 168	14 ,, 182	14 ,, 196	14 ,, 210
15 ,, 135	15 ,, 150	15 ,, 165	15 ,, 180	15 ,, 195	15 ,, 210	15 ,, 225
16 ,, 144	16 ,, 160	16 ,, 176	16 ,, 192	16 ,, 208	16 ,, 224	16 ,, 240
17 ,, 153	17 ,, 170	17 ,, 187	17 ,, 204	17 ,, 221	17 ,, 238	17 ,, 255
18 ,, 162	18 ,, 180	18 ,, 198	18 ,, 216	18 ,, 234	18 ,, 252	18 ,, 270
19 ,, 171	19 ,, 190	19 ,, 209	19 ,, 228	19 ,, 247	19 ,, 266	19 ,, 285
20 ,, 180	20 ,, 200	20 ,, 220	20 ,, 240	20 ,, 260	20 ,, 280	20 ,, 300

Multiplication Tables cont.

16 times	17 times	18 times	19 times	20 times	24 times	25 times
1 are 16	1 are 17	1 are 18	1 are 19	1 are 20	1 are 24	1 are 25
2 ,, 32	2 ,, 34	2 ,, 36	2 ,, 38	2 ,, 40	2 ,, 48	2 ,, 50
3 ,, 48	3 ,, 51	3 ,, 54	3 ,, 57	3 ,, 60	3 ,, 72	3 ,, 75
4 ,, 64	4 ,, 68	4 ,, 72	4 ,, 76	4 ,, 80	4 ,, 96	4 ,, 100
5 ,, 80	5 ,, 85	5 ,, 90	5 ,, 95	5 ,, 100	5 ,, 120	5 ,, 125
6 ,, 96	6 ,, 102	6 ,, 108	6 ,, 114	6 ,, 120	6 ,, 144	6 ,, 150
7 ,, 112	7 ,, 119	7 ,, 126	7 ,, 133	7 ,, 140	7 ,, 168	7 ,, 175
8 ,, 128	8 ,, 136	8 ,, 144	8 ,, 152	8 ,, 160	8 ,, 192	8 ,, 200
9 ,, 144	9 ,, 153	9 ,, 162	9 ,, 171	9 ,, 180	9 ,, 216	9 ,, 225
10 ,, 160	10 ,, 170	10 ,, 180	10 ,, 190	10 ,, 200	10 ,, 240	10 ,, 250
11 ,, 176	11 ,, 187	11 ,, 198	11 ,, 209	11 ,, 220	11 ,, 264	11 ,, 275
12 ,, 192	12 ,, 204	12 ,, 216	12 ,, 228	12 ,, 240	12 ,, 288	12 ,, 300
13 ,, 208	13 ,, 221	13 ,, 234	13 ,, 247	13 ,, 260	13 ,, 312	13 ,, 325
14 ,, 224	14 ,, 238	14 ,, 252	14 ,, 266	14 ,, 280	14 ,, 336	14 ,, 350
15 ,, 240	15 ,, 255	15 ,, 270	15 ,, 285	15 ,, 300	15 ,, 360	15 ,, 375
16 ,, 256	16 ,, 272	16 ,, 288	16 ,, 304	16 ,, 320	16 ,, 384	16 ,, 400
17 ,, 272	17 ,, 289	17 ,, 306	17 ,, 323	17 ,, 340	17 ,, 408	17 ,, 425
18 ,, 288	18 ,, 306	18 ,, 324	18 ,, 342	18 ,, 360	18 ,, 432	18 ,, 450
19 ,, 304	19 ,, 323	19 ,, 342	19 ,, 361	19 ,, 380	19 ,, 456	19 ,, 475
20 ,, 320	20 ,, 340	20 ,, 360	20 ,, 380	20 ,, 400	20 ,, 480	20 ,, 500

Compressed Multiplication and Division Table

	Col.	Col.	Col.	Col.	Col.	Col.	Col.	Col.	Col.	Col.	Col.	Col.	Col.
Row **1**	**2**	**3**	**4**	**5**	**6**	**7**	**8**	**9**	**10**	**11**	**12**	**13**	**14**
,, **2**	4	6	8	10	12	14	16	18	20	22	24	26	28
,, **3**	6	9	12	15	18	21	24	27	30	33	36	39	42
,, **4**	8	12	16	20	24	28	32	36	40	44	48	52	56
,, **5**	10	15	20	25	30	35	40	45	50	55	60	65	70
,, **6**	12	18	24	30	36	42	48	54	60	66	72	78	84
,, **7**	14	21	28	35	42	49	56	63	70	77	84	91	98
,, **8**	16	24	32	40	48	56	64	72	80	88	96	104	112
,, **9**	18	27	36	45	54	63	72	81	90	99	108	117	126
,, **10**	20	30	40	50	60	70	80	90	100	110	120	130	140
,, **11**	22	33	44	55	66	77	88	99	110	121	132	143	154
,, **12**	24	36	48	60	72	84	96	108	120	132	144	156	168
,, **13**	26	39	52	65	78	91	104	117	130	143	156	169	182
,, **14**	28	42	56	70	84	98	112	126	140	154	168	182	196

This table gives the product of any two factors not greater than **14** in condensed form. It can be used for either multiplication or division in the same way as can the table on page 5 for addition or subtraction. E.g., to multiply 9 by 13. Run the finger along row 9 until you come to the number in that row which is in column 13. This 117 is the required product.

Or to divide 154 by 11. Run your finger along row 11 until it comes to the number 154; then the figure at the top of the column in which it stands—i.e., 14—is the required quotient. Should the dividend not be exactly divisible by the divisor, stop at that figure next below the amount of your dividend in the row of the divisor. The top of its column gives the quotient, and the difference between the number in the column and your dividend gives the remainder. E.g., 91 ÷ 8. The next lower number in the row is 88, giving quotient 11 and remainder 3.

Mathematical Signs and Symbols

$+$	Plus, the sign of addition
$-$	Minus, the sign of subtraction
$\times$	The sign of multiplication
$\div$	The sign of division
$:$	Is to
$::$	As
$:$	Is to

(Is to, As, Is to: The signs of proportion. Thus $3:6::4:8$)

$\because$	Because
$\therefore$	Therefore
$=$	Equals, the sign of equality
$>$	Greater than
$<$	Less than
$\sqrt{}$	Square root
$\sqrt[3]{}$	Cube root. $\sqrt[4]{}$ Fourth root. $\sqrt[5]{}$ Fifth root, etc.
() [] {}	Indicate that the figures enclosed are to be taken together. Thus $10\times(7+4)$; $8-[9\div3]$; $30\times\left\{\frac{7\times3}{4-2}\right\}$
$^{\circ}\ '\ ''$	Degrees, minutes, seconds, (angular measure). Thus $25^{\circ}\,14'\,10''$ represents 25 degrees, 14 minutes, 10 seconds
$'\ ''$	Feet, inches. Thus $9'\,10'' = 9$ feet 10 inches
∞	Infinity

Signs used in geometry

$\perp$	Perpendicular to	∟	Right angle
$\parallel$	Parallel to	□	Square
○	Circle	▭	Rectangle
$\angle$	Angle	$\triangle$	Triangle

Simple Rules in Mental Arithmetic

To Multiply by	5	Add a zero and divide by		2.
,, ,,	25	,, 2 zeros	,,	4.
,, ,,	50	,, 2 ,,	,,	2.
,, ,,	250	,, 3 ,,	,,	4.
,, ,,	500	,, 3 ,,	,,	2.
,, ,,	99	,, 2 zeros to the multiplicand, and		

from this number take away the original multiplicand.

To Multiply by 999 Add 3 zeros, and from this number take away the original multiplicand.

To Divide by 5 Multiply by 2, and place the decimal point before the last figure in the product.

To Divide by 25 Multiply by 4, and place the decimal point before the last two figures.

To Divide by 125 Multiply by 8, and place the decimal point before the last three figures.

A Number is exactly Divisible

By 2 if the unit digit is an even number.
By 3 if the sum of the digits is exactly divisible by 3.
By 4 if the number formed by the last two digits is exactly divisible by 4.
By 5 if the last digit is 5 or a zero.
By 8 if the number formed by the last three digits is exactly divisible by 8.
By 9 if the sum of the digits is divisible by 9.
By 11 if the sum of the digits in the odd places is equal to, or differs by a multiple of 11 from, the sum of the digits in the even places.
By 25 if the last two digits are zeros, or exactly divisible by 25.

Area, Volume and Circumference

Note: In the following formulae $\pi = 3{\cdot}14159$ or nearly $3\frac{1}{7}$; l = length of side; h = perpendicular height from base; r = radius.

Area Surface is measured in square units, for example square inches (in^2) or square metres (m^2). A square inch is a square with each side equal to one inch. There are 3 feet in a yard but there are (3×3) or 9 ft^2 in a square yard.

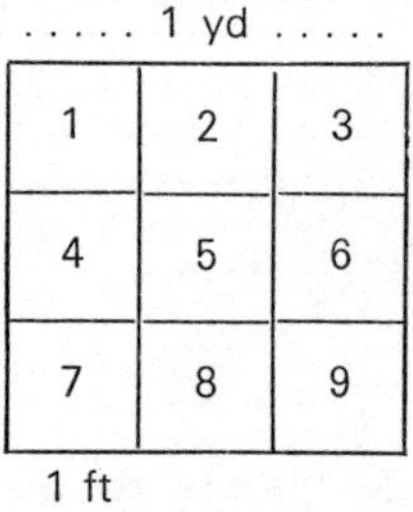

. . . . . 1 yd . . . . .

1	2	3
4	5	6
7	8	9

1 ft

Area of square $= l \times l = l^2$

,, rectangle = length × breadth

,, parallelogram = length of one side × the perpendicular distance between that side and the one parallel to it

,, triangle $= \text{base} \times \frac{1}{2}h$

,, circle $= \pi r^2$

,, surface of sphere $= 4\pi r^2$

Volume Volume is measured in cubic units, for example cubic feet (ft^3) or cubic centimetres (cm^3). A cubic foot is a cube with each edge equal to one foot. There are 3 feet in a yard and $(3 \times 3 \times 3)$ or 27 ft^3 in a cubic yard.

In the diagram each small block represents a cubic foot; it can be seen that it requires 27 of these blocks to make a cubic yard, as there are three distinct layers, each containing 9 (3 rows of 3) blocks.

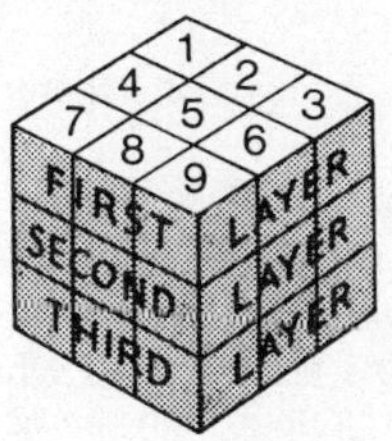

Volume of cube $= l \times l \times l = l^3$

" sphere $= \frac{4}{3}\pi r^3$

" cylinder $= \pi r^2 h$

" cone $= \frac{1}{3}$ that of cylinder on the same base and of the same height

$= \dfrac{\pi r^2 h}{3}$

Circumference of circle $= 2\pi r$.

Fractions

A **fraction** is a numerical quantity that is not a whole number. There are two kinds, **vulgar fractions,** commonly referred to as **fractions** and **decimal fractions.** commonly referred to as **decimals.**

Vulgar fractions are expressed by two numbers placed one above the other and separated by a line. The lower number is the **denominator** and the upper number is the **numerator**. The denominator shows into how many equal parts the whole is divided up and the numerator shows how many of these parts are taken. Thus $\frac{3}{5}$ of 20 means that twenty is divided into five equal parts and three of these are taken.

e.g. $20 \div 5 = 4$, $3 \times 4 = 12$, therefore $\frac{3}{5}$ of $20 = 12$.

Proper fractions are fractions which are less than 1, that is to say where the numerator is less than the denominator.

Improper fractions are fractions which are greater than 1; the numerator is greater than the denominator.

e.g. $\frac{7}{6} = 1\frac{1}{6}$, $\frac{17}{4} = 4\frac{1}{4}$.

Decimal fractions are indicated by the presence of a decimal point. This point is placed to the right of the unit of a number and figures placed to the right of the point represent tenths, hundredths, thousandths, etc. of a unit according to their place; those next to the point representing tenths of a unit, the next hundredths and so on.

Thus $1{\cdot}2 = 1 + \frac{2}{10}$, $0{\cdot}23 = \frac{2}{10} + \frac{3}{100} = \frac{23}{100}$.

The Decimal System makes it easy to calculate and forms the basis of the Metric System of weights and measures.

Percentage equivalents of fractions are formed by multiplying the fraction by 100.

e.g. $\frac{2}{5} = (\frac{2}{5} \times 100)\% = 40\%$,
$0{\cdot}25 = (0{\cdot}25 \times 100)\% = 25\%$.

Notation

Notation is the representation of numbers by figures.

Numeration is the conversion of numerals from figures into words.

The figures generally used today are of Arabic origin. Roman numerals are sometimes used for numbering. For example, chapters in books and the dial-plates on watches and clocks are often numbered with Roman numerals.

Arabic numbers The Arabic figures are 1, one; 2, two; 3, three; 4, four; 5, five; 6, six; 7, seven; 8, eight; 9, nine. All our numbers are formed by these and zero, 0.

Roman numbers These are expressed using seven letters of the alphabet: I, one; V, five; X, ten; L, fifty; C, a hundred; D, five hundred; M, a thousand. If one of the symbols is preceded by one of less value the number represented is equal to the difference of the values, e.g. IX = 10 − 1 = 9. The number represented by a Roman numeral can be determined by adding together the values of any pairs formed by a symbol preceding a symbol of greater value and the values of the remaining individual symbols.

e.g. MCMLXXIX = M(CM)LXX(IX)
= 1000 + 900 + 50 + 10 + 10 + 9 = 1979

Table of Roman Numerals

1 = I	12 = XII	35 = XXXV	100 = C
2 = II	13 = XIII	40 = XL	200 = CC
3 = III	14 = XIV	45 = XLV	300 = CCC
4 = IV or IIII	15 = XV	50 = L	400 = CD
5 = V	16 = XVI	55 = LV	500 = D
6 = VI	17 = XVII	60 = LX	600 = DC
7 = VII	18 = XVIII	65 = LXV	700 = DCC
8 = VIII	19 = XIX	70 = LXX	800 = DCCC
9 = IX	20 = XX	75 = LXXV	900 = CM
10 = X	25 = XXV	80 = LXXX	1000 = M
11 = XI	30 = XXX	90 = XC	2000 = MM

Number Systems

Denary System This system is the number system used for practically all purposes. It uses the ten digits 0 to 9. The position of a figure in a several digit number determines the value it represents. The value is multiplied by **10** at each removal to the left. If the number 7654321 is analysed we have:

The 1 = 1 unit = 1 × 1 = ...1
,, 2 = 2 tens = 2 × 10 = ...20
,, 3 = 3 hundreds = 3 × 100 = ...300
,, 4 = 4 thousands = 4 × 1,000 = ...4,000
,, 5 = 5 tens of thousands = 5 × 10,000 = ...50,000
,, 6 = 6 hundreds of thousands = 6 × 100,000 = ...600,000
,, 7 = 7 millions = 7 × 1,000,000 = ...7,000,000

Now add, and we get the original number 7,654,321

Binary System This system has gained importance with the advent of computers. It uses only the two digits 0 and 1, and is used to represent numbers in a computer because the 0 and 1 can be indicated by the absence or presence of a signal, or by a switch being off or on. As in the Denary System, the position of a digit determines the value it represents but the value is multiplied by **2** at each removal to the left. If the number 110101 is analysed we have, starting from the right:

The first digit 1 = 1 × 1 = ... 1
,, second ,, 0 = 0 × 2 = ... 0
,, third ,, 1 = 1 × 2 × 2 = ... 4
,, fourth ,, 0 = 0 × 2 × 2 × 2 = ... 0
,, fifth ,, 1 = 1 × 2 × 2 × 2 × 2 = ... 16
,, sixth ,, 1 = 1 × 2 × 2 × 2 × 2 × 2 = ... 32
Now add and we get the denary number represented by the binary number 110101 53

Table of the first 16 binary numbers

1 = 1	101 = 5	1001 = 9	1101 = 13
10 = 2	110 = 6	1010 = 10	1110 = 14
11 = 3	111 = 7	1011 = 11	1111 = 15
100 = 4	1000 = 8	1100 = 12	10000 = 16

CURRENCY AND COMMERCE

The **pound** sterling is the major unit of currency of the United Kingdom. It is divided into a hundred **pence.**

Coinage

50p, 10p and 5p are cupro-nickel coins.
2p, 1p and $\frac{1}{2}$p are decimal bronze coins.

There are Bank of England notes to the value of £20, £10, £5 and £1.

Calculations in Decimal Currency

A decimal currency makes the calculations involved in cash transactions very simple.

Examples:

Cost of 8 articles costing 2p each $= (8 \times 2)$p $=$ 16p
$=$ £0·16

Cost of 6 articles costing 25p each $= (6 \times 25)$p
$=$ 150p $=$ £1·50

Cost of 3 articles costing £1·30 each $=$ £$(3 \times 1{\cdot}30)$
$=$ £3·90

Fractional Parts of £1

and the Percentage Equivalents of the Fractions

Parts of £1 (rounded)		**Percentage equivalents** (rounded)
50p	$\frac{1}{2}$	50 %
33$\frac{1}{2}$p	$\frac{1}{3}$	33 %
66$\frac{1}{2}$p	$\frac{2}{3}$	67 %
25p	$\frac{1}{4}$	25 %

Parts of £1 (rounded)		Percentage equivalents (rounded)
75p	$\frac{3}{4}$	75%
20p	$\frac{1}{5}$	20%
40p	$\frac{2}{5}$	40%
60p	$\frac{3}{5}$	60%
80p	$\frac{4}{5}$	80%
$12\frac{1}{2}$p	$\frac{1}{8}$	$12\frac{1}{2}$%
10p	$\frac{1}{10}$	10%
5p	$\frac{1}{20}$	5%
$2\frac{1}{2}$p	$\frac{1}{40}$	$2\frac{1}{2}$%
2p	$\frac{1}{50}$	2%
1p	$\frac{1}{100}$	1%

Simple and Compound Interest

The term **Interest** in commerce means the money paid for the use of money lent.

Simple Interest is calculated on the original loan. The amount originally lent is called the **principal**. The simple interest which accumulates can be calculated from the formula

$$I = \frac{P \times R \times T}{100}$$

where P is the principal, R is the rate per cent and T the time.

The sum to which the principal will grow is given by the formula

$$S = P\left(1 + \frac{R \times T}{100}\right)$$

Compound Interest is the term used when interest is paid not only on the principal but also on the interest as it accumulates. The sum to which the principal will grow is given by the formula

$$S = P(1+i)^n$$

where i is the periodic interest and n the number of periods.

Time at which Money Doubles Itself

Interest Rate	**Simple Interest**			**Compound Interest**			
1%	100 years			70 years			
2%	50	,,		35	,,	1 day	
3%	33	,,	4 months	23	,,	164 days	
4%	25	,,		17	,,	246	,,
5%	20	,,		14	,,	75	,,
6%	16	,,	8 months	11	,,	327	,,
7%	14	,,	104 days	10	,,	89	,,
8%	12	,,	6 months	9	,,	2	,,
9%	11	,,	40 days	8	,,	16	,,
10%	10	,,		7	,,	100	,,

Simple Interest

Table showing the sum to which £100 will grow

Interest rate (per cent per annum)

Years	$\frac{1}{8}$% £	$\frac{1}{8}$% p	$\frac{1}{4}$% £	$\frac{1}{4}$% p	$\frac{1}{2}$% £	$\frac{1}{2}$% p	1% £	1% p	2% £	2% p	3% £	3% p	4% £	4% p
1	100	12½	100	25	100	50	101	00	102	00	103	00	104	00
2	100	25	100	50	101	00	102	00	104	00	106	00	108	00
3	100	37½	100	75	101	50	103	00	106	00	109	00	112	00
4	100	50	101	00	102	00	104	00	108	00	112	00	116	00
5	100	62½	101	25	102	50	105	00	110	00	115	00	120	00
6	100	75	101	50	103	00	106	00	112	00	118	00	124	00
7	100	87½	101	75	103	50	107	00	114	00	121	00	128	00
8	101	00	102	00	104	00	108	00	116	00	124	00	132	00
9	101	12½	102	25	104	50	109	00	118	00	127	00	136	00
10	101	25	102	50	105	00	110	00	120	00	130	00	140	00
15	101	87½	103	75	107	50	115	00	130	00	145	00	160	00
20	102	50	105	00	110	00	120	00	140	00	160	00	180	00
25	103	12½	106	25	112	50	125	00	150	00	175	00	200	00
30	103	75	107	50	115	00	130	00	160	00	190	00	220	00
40	105	00	110	00	120	00	140	00	180	00	220	00	260	00
50	106	25	112	50	125	00	150	00	200	00	250	00	300	00

Simple Interest cont.

Years	5% £	5% p	6% £	6% p	7% £	7% p	8% £	8% p	9% £	9% p	10% £	10% p
1	105	00	106	00	107	00	108	00	109	00	110	00
2	110	00	112	00	114	00	116	00	118	00	120	00
3	115	00	118	00	121	00	124	00	127	00	130	00
4	120	00	124	00	128	00	132	00	136	00	140	00
5	125	00	130	00	135	00	140	00	145	00	150	00
6	130	00	136	00	142	00	148	00	154	00	160	00
7	135	00	142	00	149	00	156	00	163	00	170	00
8	140	00	148	00	156	00	164	00	172	00	180	00
9	145	00	154	00	163	00	172	00	181	00	190	00
10	150	00	160	00	170	00	180	00	190	00	200	00
15	175	00	190	00	205	00	220	00	235	00	250	00
20	200	00	220	00	240	00	260	00	280	00	300	00
25	225	00	250	00	275	00	300	00	325	00	350	00
30	250	00	280	00	310	00	340	00	370	00	400	00
40	300	00	340	00	380	00	420	00	460	00	500	00
50	350	00	400	00	450	00	500	00	550	00	600	00

Simple Interest cont.

Years	11%		12%		13%		14%		15%	
	£	p	£	p	£	p	£	p	£	p
1	111	00	112	00	113	00	114	00	115	00
2	122	00	124	00	126	00	128	00	130	00
3	133	00	136	00	139	00	142	00	145	00
4	144	00	148	00	152	00	156	00	160	00
5	155	00	160	00	165	00	170	00	175	00
6	166	00	172	00	178	00	184	00	190	00
7	177	00	184	00	191	00	198	00	205	00
8	188	00	196	00	204	00	212	00	220	00
9	199	00	208	00	217	00	226	00	235	00
10	210	00	220	00	230	00	240	00	250	00
15	265	00	280	00	295	00	310	00	325	00
20	320	00	340	00	360	00	380	00	400	00
25	375	00	400	00	425	00	450	00	475	00
30	430	00	460	00	490	00	520	00	550	00
40	540	00	580	00	620	00	660	00	700	00
50	650	00	700	0	750	00	800	00	850	00

Compound Interest

Table showing the sum to which £100 will grow when the interest is paid yearly.
(Correct to the nearest penny.)

	1%		2%		3%		4%		5%		6%		7%		8%	
Years	£	p	£	p	£	p	£	p	£	p	£	p	£	p	£	p
1	101	00	102	00	103	00	104	00	105	00	106	00	107	00	108	00
2	102	01	104	04	106	09	108	16	110	25	112	36	114	49	116	64
3	103	03	106	12	109	27	112	49	115	76	119	10	122	50	125	97
4	104	06	108	24	112	55	116	99	121	55	126	25	131	08	136	05
5	105	10	110	41	115	93	121	67	127	63	133	82	140	26	146	93
6	106	15	112	62	119	41	126	53	134	01	141	85	150	07	158	69
7	107	21	114	87	122	99	131	59	140	71	150	36	160	58	171	38
8	108	29	117	17	126	68	136	86	147	75	159	39	171	82	185	09
9	109	37	119	51	130	48	142	33	155	13	168	95	183	85	199	90
10	110	46	121	90	134	39	148	02	162	89	179	09	196	72	215	89
15	116	10	134	59	155	80	180	09	207	89	239	66	275	90	317	22
20	122	02	148	60	180	61	219	11	265	33	320	71	386	97	466	10
25	128	24	164	06	209	38	266	58	338	64	429	19	542	74	684	85
30	134	79	181	14	242	73	324	34	432	19	574	35	761	23	1006	27
40	148	89	220	80	326	20	480	10	704	00	1028	57	1497	45	2172	45
50	164	46	269	16	438	39	710	67	1146	74	1842	02	2945	70	4690	16

Wages Table for the Year, Quarter, Calendar Month, Week and Day

Yearly £	p	Quarterly £	p	Monthly £	p	Weekly £	p	Daily £	p
0	50	0	12½	0	04	0	01	0	00
1	00	0	25	0	08½	0	02	0	00½
2	00	0	50	0	16½	0	04	0	00½
3	00	0	75	0	25	0	06	0	01
4	00	1	00	0	33½	0	07½	0	01
5	00	1	25	0	41½	0	09½	0	01½
6	00	1	50	0	50	0	11½	0	01½
7	00	1	75	0	58½	0	13½	0	02
8	00	2	00	0	66½	0	15½	0	02
9	00	2	25	0	75	0	17½	0	02½
10	00	2	50	0	83½	0	19	0	02½
20	00	5	00	1	66½	0	38½	0	05½
30	00	7	50	2	50	0	57½	0	08
40	00	10	00	3	33½	0	77	0	11
50	00	12	50	4	16½	0	96	0	13½
60	00	15	00	5	00	1	15½	0	16½
70	00	17	50	5	83½	1	34½	0	19
80	00	20	00	6	66½	1	54	0	22
90	00	22	50	7	50	1	73	0	24½
100	00	25	00	8	33½	1	92½	0	27½
200	00	50	00	16	66½	3	84½	0	55
300	00	75	00	25	00	5	77	0	82
400	00	100	00	33	33½	7	69	1	09½
500	00	125	00	41	66½	9	61½	1	37
600	00	150	00	50	00	11	54	1	64½
700	00	175	00	58	33½	13	46	1	92
800	00	200	00	66	66½	15	38½	2	19
900	00	225	00	75	00	17	31	2	46½
1000	00	250	00	83	33½	19	23	2	74

(All entries correct to the nearest ½p.)

Any wage can be accommodated by this table by breaking the amount down into halves of units, units, tens, hundreds and thousands, e.g.

$$£2345{\cdot}50 = 2 \times £1000 + £300 + £40 + £5 + £0{\cdot}50$$

So £2345·50 yearly

$$= (2 \times £19{\cdot}23 + £5{\cdot}77 + £0{\cdot}77 + £0{\cdot}09\tfrac{1}{2} + £0{\cdot}01) \quad \text{weekly}$$

$$= £45{\cdot}10\tfrac{1}{2} \quad \text{weekly}$$

Currencies of the World

Country	Monetary Unit	Value of unit in British Currency (Dec. 1973)
Australia	Dollar = 100 Cents	£0·647
Austria	Schilling = 100 Groschen	£0·022
Belgium	Franc = 100 Centimes	£0·011
Canada	Dollar = 100 Cents	£0·435
Denmark	Krone = 100 Öre	£0·069
France	Franc = 100 Centimes	£0·094
Germany	Deutsche Mark = 100 Pfennig	£0·163
Holland	Guilder = 100 Cents	£0·155
Italy	Lira = 100 Centesimi	£0·001
Japan	Yen	£0·002
Norway	Krone = 100 Öre	£0·076
Portugal	Escudo = 100 Centavos	£0·017
Spain	Peseta = 100 Céntimos	£0·008
Sweden	Krona = 100 Öre	£0·094
Switzerland	Franc = 100 Centimes	£0·135
U.S.A.	Dollar = 100 Cents	£0·432

Commercial Symbols

£	pound sterling
$	dollar
p	penny
%	per cent
c/o	care of
a/c	account
@	at

WEIGHTS AND MEASURES

The **Imperial System** of measurement, based on the yard for length and the pound for weight, has been used in the United Kingdom for many centuries.

The **Metric System** was founded by the French during the French Revolution.

The **Système International d'Unités** (International System of Units) is the modern form of the metric system finally agreed at an international conference in 1960. The international symbol for this system is **SI**. SI is the result of the rationalisation of the whole structure of metric derived units. The United Kingdom decided to adopt SI as the primary system of measures and weights because almost 90% of the world's population live in countries that use SI, or are committed to use it.

THE INTERNATIONAL SYSTEM OF UNITS

SI is based on six **base-units**:

Quantity	Unit	Symbol
length	metre	m
mass	kilogramme	kg
time	second	s
electric current	ampere	A
temperature	kelvin	K
luminous intensity	candela	cd

The SI units for plane angle and solid angle, the radian (rad) and steradian (sr) respectively, are called **supplementary units**.

The **derived units** are stated in terms of the base-units, e.g., velocity is expressed in metres per second (m/s). Some derived units are given special names, e.g., the watt (W).

Note. By international agreement the United Kingdom, Europe and most other countries use the spellings *metre* and *kilogramme*. In North America the spellings are *meter* and *kilogram*.

Larger or smaller quantities are formed by multiplying or dividing the SI units by 10, 100, 1000 and so on. The names of these quantities are formed by means of prefixes to the principal units.

prefix		symbol
mega	— a million times	**M**
kilo	— a thousand times	**k**
hecto *	— a hundred times	**h**
deca *	— ten times	**da**
deci *	— a tenth part of	**d**
centi *	— a hundredth part of	**c**
milli	— a thousandth part of	**m**
micro	— a millionth part of	**µ**

* *The use of these prefixes is avoided unless absolutely necessary.*

Examples:

1 millimetre (mm) = $\frac{1}{10}$ centimetre (cm) = $\frac{1}{1000}$ metre (m)
1 megawatt (MW) = 1000 kilowatts (kW) = 1 000 000 watts (W)

Length

Metre (m) is the SI base-unit for length.

The metre was intended to be one ten-millionth part of the distance from the North Pole to the Equator at sea-level through Paris.

1 m = 1·094 yd = 39·37 in	1 in = 25·4 mm
	1 ft = 0·304 8 m
	= 30·48 cm
1 mm = 0·039 4 in	1 yd = 0·914 4 m
1 km = 0·621 4 mile = 1093·6 yd	1 mile = 1·609 km

Examples:

Imperial Measure	**SI Units** (approx.)
Man's height 5 ft 11 in	180 cm or 1·8 m
Diameter of a 6-in plate	15 cm
Length of a cricket pitch: 22 yd = 1 chain	20·1 m
Distance from Dover to Calais: 22 miles	35·4 km
Distance from London to Manchester: 184 miles	296 km
Distance from London to Frankfurt: 406 miles	653 km
Distance from London to New York: 3442 miles	5539 km
Distance from the Earth to the Moon: 238 840 miles	384 380 km

Approximate Conversions:

Metres into feet: multiply by $3^1/_4$

Metres into yards: add one tenth

Kilometres into miles: multiply by $^5/_8$

Millimetres into inches: divide by 25

Yards into metres: deduct one-tenth

Feet into metres: multiply by $^3/_{10}$

Miles into kilometres: multiply by $^8/_5$

Inches into millimetres: multiply by 25

Area

Square metre (m^2) is the derived SI unit for area.

There are also two special units which can be used.

Hectare (ha) = 10 000 m^2

Are (a) = 100 m^2

1 m^2 = 1·196 yd^2

1 ha = 2·471 acres
1 a = 1076 ft^2
1 km^2 = 247·1 acres = 0·386 1 sq. mile

1 in^2 = 645·16 mm^2
1 ft^2 = 0·092 9 m^2 = 929·03 cm^2
1 yd^2 = 0·836 1 m^2
1 acre = 4046·86 m^2 = 0·404 7 ha
1 sq. mile = 2·590 km^2 = 258·999 ha

Examples:

Imperial Measure	**SI Units** (approx.)
Area of a football pitch: 120 yd × 80 yd = 9600 yd^2	8027 m^2
Area of a doubles tennis court: 78 ft × 36 ft = 2808 ft^2	261 m^2
Area of an average carpet: 4 yd × 3 yd = 12 yd^2	10 m^2
Area of an average house-plot: $\frac{1}{3}$ acre	0·13 ha
Surface area of a 6-in diameter plate: 28·3 in^2	183 cm^2

Approximate Conversions:

Square metres into square yards: add one-fifth

Hectares into acres: multiply by $\frac{5}{2}$

Acres into hectares: multiply by $\frac{2}{5}$

Volume

Cubic metre (m^3) is the derived SI unit for volume.

1 m^3 = 1·308 0 yd^3

1 dm^3 = 0·035 3 ft^3
1 cm^3 = 0·061 0 in^3

1 in^3 = 16·387 cm^3
1 ft^3 = 0·028 3 m^3
1 yd^3 = 0·764 6 m^3

Examples:

Imperial Measure	**SI Units** (approx.)
Volume of a 3-lb biscuit tin: 9 in × 9½ in × 5 in = 427·5 in^3	7005 cm^3
Volume of an average room: 12 ft × 14 ft × 10 ft = 1680 ft^3	47·6 m^3

Approximate Conversions:

Cubic metres into cubic yards: add one-third

Cubic yards into cubic metres: deduct one-third

Capacity

(For volumes of Fluids)

Litre (l) is the SI unit used for general measures of capacity but not for scientific measurements, when the cubic metre, cubic millimetre, etc., are used.

1 l = 1 dm^3

There are less widely used multiples and sub-multiples, in particular the **hectolitre** (hl) and **millilitre** (ml).

1 l = 0·220 gal = 1·760 pt = 61·026 in^3

1 hl = 22·00 gal

1 gal = 4·546 l
1 qt = 1·136 l
1 pt = 0·568 l
1 gill = 142 ml
1 fl. oz = 28·4 ml

continued

Examples:

Imperial Measure	**SI Units** (approx.)
1 British teaspoon holds approx. $\frac{1}{25}$ gill = $\frac{1}{5}$ fl. oz	5·6 ml
4 gallons of petrol	18 l
Pint of milk, beer, paint, etc.	0·57 l
40-gal drum of petrol or oil	1·8 hl

Approximate Conversions:

Litres into pints:
add three-quarters

Pints into litres:
multiply by $\frac{3}{5}$

Weight

Kilogramme (kg) is the SI base-unit for weight.

The gramme was intended to be the mass of one cubic centimetre (one millilitre) of water at 0°C.

1 tonne (t) (metric ton) = 1000 kg
1 metric carat = 0·2 g

The metric carat is used for commercial transactions in diamonds, fine pearls and precious stones.

1 kg = 2·205 lb

1 g = 0·035 oz
1 t = 0·984 ton

1 oz = 28·350 g
1 lb = 454 g
1 stone = 6·350 kg
1 cwt = 50·802 kg
1 ton = 1·016 t

Examples:

Imperial Measure	**SI Units** (approx.)
$\frac{1}{2}$ lb butter	0·23 kg or 227 g
3 lb flour	1·4 kg
5 cwt coal	254 kg
A birth weight of 8$\frac{1}{2}$ lb	3856 g or 3·9 kg
A man weighing 13$\frac{1}{2}$ stone = 189 lb	85·7 kg

continued

Approximate Conversions:

Kilogrammes into pounds: add a tenth and multiply by 2
Pounds into kilogrammes: deduct a tenth and divide by 2
Ounces into grammes: multiply by 28
Hundredweights into kilogrammes: multiply by 50
The tonne and ton are *nearly* equal.

Domestic Metric Measures

Weight

The basic unit for recipes is 25 g (1 oz = 28·35g).

Volume

The millilitre is used for all measures of capacity up to 1 litre.

Equipment

Measuring Spoons 1·25, 2·5, 5, 10, 15, 20 ml
(A 5 ml spoon is **approximately** the same as a teaspoon.)

Measuring Cups 50 ml ($\frac{1}{6}$)
75 ml ($\frac{1}{4}$)
100 ml ($\frac{1}{3}$)
150 ml ($\frac{1}{2}$)
300 ml (full)
(50 ml cup holds 50 g sugar; 75 ml cup holds 50 g flour.)

Measuring Jugs 1 litre, marked at intervals, showing 7·5 dl, 5 dl, 1·25 dl and 1 dl.
$\frac{1}{2}$ litre, marked at intervals, showing 2·5 dl, 1·25 dl, 1 dl and smaller intervals if required.

Kitchen Scales Calibrated in 25 g units (differing from scales used by the trade which are calibrated with intervals of 10 g or 20 g).

Baking Tins 6 in (15·2 cm) becomes 15 cm
8 in (20·3 cm) becomes 20 cm

Temperatures Ovens are calibrated in Centigrade (formerly Fahrenheit was used). The main divisions are at 50 C intervals—ranging from 100 C to 250 C—with intermediate markings at 10 C intervals.

Temperature Equivalents for Oven Thermostat Markings

° Fahrenheit Scale	° Centigrade Scale
150° F	70° C
175° F	80° C
200° F	100° C
225° F	110° C
250° F	130° C
275° F	140° C
300° F	150° C
325° F	170° C
350° F	180° C
375° F	190° C
400° F	200° C
425° F	220° C
450° F	230° C
475° F	240° C
500° F	250° C
525° F	270° C
550° F	290° C

Food Packaging

Dry Foods Breakfast cereals, bread, tea, cocoa, coffee, honey, jam, marmalade, syrup, salt, sugar, flour, rice, pastas, butter, fats, if pre-packed, must be sold in weights chosen from a specified series of weights. At the time of writing a range of metric sizes has been recommended to replace the series of Imperial weights, namely:

grammes: 25, 50, 125, 250, 500
kilogrammes: 1, $1\frac{1}{2}$, 2

Temperature

Kelvin (K) is the SI base-unit for temperature.

Degrees Kelvin, also known as degrees **Absolute**, are mainly used for scientific measurements. The customary international unit is the degree **Celsius** which is known as **Centigrade** (°C) in the United Kingdom. (In France, Centigrade is a unit of angular measure and the name Celsius is used for temperature.) The zero of the Centigrade scale is the temperature of the ice-point (273·16°K). The units of the Centigrade and Kelvin temperature interval are the same. Another unit used in the United Kingdom is the degree **Fahrenheit** (°F).

	°C	°F
Ice point	0	32
Boiling point of water under standard pressure	100	212

To convert °F to °C deduct 32 and multiply by $\frac{5}{9}$.
To convert °C to °F multiply by $\frac{9}{5}$ and add 32.

Examples:

	°F	°C
Temperature on a very frosty day	12	−11
Temperature on a hot summer day	86	30
Normal body temperature	98·4	37

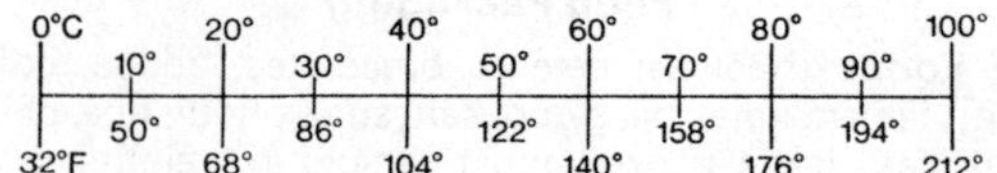

Angular Measure

Radian (rad) is the SI unit for angular measure. It is a supplementary SI Unit.

A more commonly used unit for angular measure is the **degree** which is subdivided into **minutes** and **seconds.**

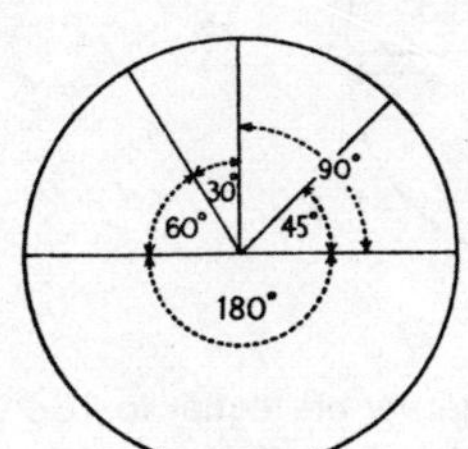

60 seconds (″) = 1 minute
60 minutes (′) = 1 degree
90 degrees (°) = 1 right angle
360 degrees (°) = 1 full circle

An **acute** angle is one less than 90°.

An **obtuse** angle is one between 90° and 180°.

A **reflex** angle is one between 180° and 360°.

The circumference of a circle is 3·141 59, or nearly $3\frac{1}{7}$, times its diameter. This number 3·141 59 is known as **π**, a widely used mathematical constant.

A radian is the angle between two radii of a circle which

cut off on the circumference an arc equal in length to the radius.

π rad = 180°

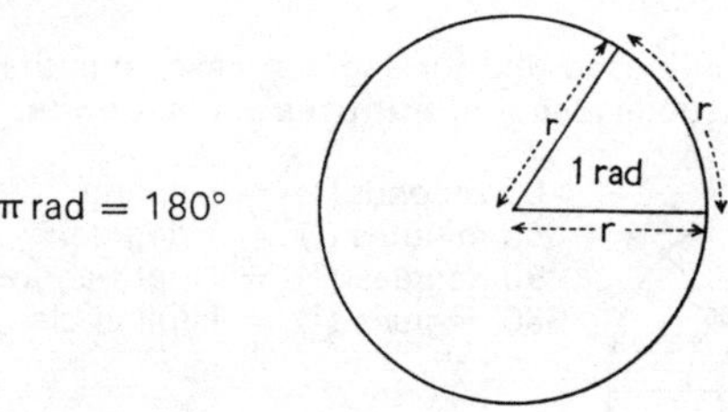

The angles of a triangle added together are equal to 180°.

Velocity

Velocity can be measured in **metres per second** (m/s) or **kilometres per hour** (km/h) in SI.

1 km/h = 0·621 mile/h
1 m/s = 3·281 ft/s

1 mile/h = 1·609 km/h
1 ft/s = 0·304 8 m/s

Examples:

Imperial Measure	**SI Units** (approx.)
30 mile/h = 44 ft/s	48 km/h = 13 m/s
50 mile/h	80 km/h
70 mile/h	113 km/h

Approximate Conversions:

km/h into mile/h:
multiply by $\frac{5}{8}$

mile/h into km/h:
multiply by $\frac{8}{5}$

Force

Newton (N) is the derived SI unit of Force. It is the force which when applied to a mass of 1 kilogramme gives it an acceleration of 1 metre per second per second.

The **dyne** is a non-SI metric unit.

1 N = 100 000 dynes

Pressure

The SI unit of pressure or stress is the **newton per square metre** (N/m^2).

1 N/m^2 = 0·000 145 lbf/in^2	1 lbf/in^2 (1 p.s.i.) = 6894·76 N/m^2
1 kN/m^2 = 0·295 inHg	1 inHg = 3383·39 N/m^2

The **millibar** (mb or mbar) and **millimetres of mercury** (mmHg) are non-SI metric units used to measure atmospheric pressure.

1 mbar = 100 N/m^2
1 mmHg = 133·322 N/m^2

Energy

Joule (J) is the SI unit which measures work and quantity of heat. It is the work done when the point of application of a force of 1 newton is displaced through a distance of 1 metre in the direction of the force.

1 J = 0·738 ft lbf	1 therm = 105·5 MJ
1 kJ = 0·278 Wh	1 kWh = 3·6 MJ
	1 Btu = 1·0554 kJ

The **calorie**, **kilo-calorie** and **Calorie** are non-SI metric units.

1 cal = 4·186 8 J

The kilo-calorie and Calorie (spelt with a capital 'C') are

equal to 1000 calories. The kilo-calorie is used for measurements of large quantities of heat. The Calorie is used for describing the energy values of foods.

Power

Watt (W) is the SI unit of power. The watt is equal to 1 joule per second.

$$1 \text{ hp} = 745{\cdot}7 \text{ W}$$

Electricity

Ampere (A) is the SI base-unit for **electric current**.

Coulomb (C) is the SI unit of **electric charge.** It is the quantity of electricity transported in 1 second by 1 ampere.

Volt. (V) is the SI unit of **electric potential**. A volt is the difference of electric potential between two points of a conducting wire carrying a constant current of 1 ampere when the power dissipated between these points is equal to 1 watt.

Ohm (Ω) is the SI unit of **electric resistance.** A resistance of 1 ohm between two points of a conductor with a constant difference of potential of 1 volt produces in the conductor a current of 1 ampere.

Farad (F) is the SI unit of **electric capacitance**. It is the capacitance of a capacitor charged to a potential of 1 volt by 1 coulomb of electricity.

Formulae:

Volts = Amperes × Ohms $(V = A \times \Omega)$
Watts = Volts × Amperes $(W = V \times A)$

THE IMPERIAL SYSTEM OF UNITS

Length

1 inch (in or ″)		
1 foot (ft or ′)	=	12 inches
1 yard (yd)	=	36 inches 3 feet
1 fathom (fm)	=	6 feet 2 yards
1 link		7·92 inches
1 rod, pole or perch	=	25 links 5½ yards
1 chain (ch)	=	100 links 22 yards 4 poles
1 furlong (fur)	=	220 yards 40 poles 10 chains
1 mile	=	1760 yards 80 chains 8 furlongs

Area

1 square inch (sq. in or in^2)		
1 square foot (sq. ft or ft^2)	=	144 (12 × 12) sq. inches
1 square yard (sq. yd or yd^2)	=	9 (3 × 3) sq. feet
1 square rod, pole or perch	=	625 (25 × 25) sq. links 30¼ (5½ × 5½) sq. yards
1 square chain	=	10 000 (100 × 100) sq. links 16 (4 × 4) sq. poles
1 rood	=	1210 sq. yards 40 sq. rods
1 acre	=	100 000 sq. links 10 sq. chains 4 roods
1 square mile	=	640 acres
1 hide of land	=	100 acres
1 barony	=	40 hides

Volume

1 cubic inch (cu. in or in³)		
1 cubic foot (cu. ft or ft³)	=	1728 (12 × 12 × 12) cu. inches
1 cubic yard (cu. yd or yd³)	=	27 (3 × 3 × 3) cu. feet

Capacity

1 fluid ounce (fl. oz)		
1 gill	=	5 fl. oz
1 pint (pt)	=	20 fl. oz 4 gills
1 quart (qt)	=	2 pints
1 gallon (gal)	=	8 pints 4 quarts
1 peck (pk)	=	2 gallons
1 bushel	=	8 gallons
1 quarter	=	8 bushels
1 chaldron	=	36 bushels

Weight

1 grain (gr)		
1 dram (dr)		
1 ounce (oz)	=	16 drams
1 pound (lb)	=	7000 grains 16 oz
1 stone	=	14 lb
1 quarter	=	28 lb 2 stone
1 cental	=	100 lb
1 hundredweight (cwt)	=	112 lb 8 stone 4 quarters
1 ton	=	2240 lb 20 cwt

SPECIALISED MEASURES

It is anticipated that by 1975 most industries and trades will have completed the changeover from the Imperial System of weights and measures to SI. At the time of writing (May 1972) information was not available to the author from all the industrial and trade federations, but details of metric standards have been included where known. As metrication becomes widespread non-SI measures will become obsolete.

Paper and Book Measures

International (ISO) Sizes

Stationery, Books and Magazines ('A' series)

	mm	in		mm	in
2A	1189 × 1682	46·81 × 66·22	A5	148 × 210	5·83 × 8·27
A0	841 × 1189	33·11 × 46·81	A6	105 × 148	4·13 × 5·83
A1	594 × 841	23·39 × 33·11	A7	74 × 105	2·91 × 4·13
A2	420 × 594	16·54 × 23·39	A8	52 × 74	2·05 × 2·91
A3	297 × 420	11·69 × 16·54	A9	37 × 52	1·46 × 2·05
A4	210 × 297	8·27 × 11·69	A10	26 × 37	1·02 × 1·46

In addition there is a series of 'B' sizes intermediate between any two adjacent sizes of the 'A' series (for posters, etc.) and of 'C' sizes (for envelopes).

Bound books

	Size of untrimmed page, mm
Metric crown octavo	192 × 126
Metric large crown octavo	204 × 132
Metric demy octavo	222 × 141
Metric royal octavo	240 × 159
A5	215 × 152·5

Note—the standard size for paperbacks is a trimmed page size of 180 mm × 110 mm.

Paper Measure

Writing and Drawing Paper		Printing Paper	
1 quire	= 24 sheets	1 ream	= 516 sheets
1 ream	= 480 sheets	1 bundle	= 2 reams
	= 20 quires	1 bale	= 5 bundles

Construction Measures

Standard Bricks

Metric	Imperial
215 mm × 102·5 mm × 65 mm	$8\frac{5}{8}$ in × $4\frac{1}{8}$ in × $2\frac{5}{8}$ in (219 mm × 104·8 mm × 66·7 mm)

Timber

1 load unhewn timber	= 40 ft³	= 1·1 m³
1 load square timber	= 50 ft³	= 1·4 m³
1 shipping ton	= 42 ft³	= 1·2 m³
1 cord	= 128 ft³	= 3·6 m³

Metric Sizes for Sawn Softwood

Lengths. The agreed stock lengths begin at 1·80 metres and rise by increments of 0·30 metres to give a series:

1·80; 2·10; 2·40; 2·70; 3·00; 3·30; 3·60; 3·90; 4·20; 4·50; 4·80; 5·10; 5·40; 5·70; 6·00; 6·30.

Cross-Sections The following table gives the cross-section sizes for sawn softwoods

Basic Metric Stock Sizes for Sawn Softwoods

Thickness, in mm	Width, mm									
	75	100	115	125	150	175	200	225	250	300
16	×	×		×	×					
19	×	×		×	×					
22	×	×		×	×					
25	×	×		×	×	×	×	×	×	×
32	×	×	×	×	×	×	×	×	×	×
38	×	×	×	×	×	×	×	×	×	×
44	×	×	×	×	×	×	×	×	×	×
50	×	×	×	×	×	×	×	×	×	×
63		×		×	×	×	×	×		
75		×		×	×	×	×	×	×	×
100		×			×		×		×	×
150					×		×			×
200							×			
250									×	
300										×

Pharmaceutical Measures

Medicine is now measured in metric units. Prescriptions are in 5 millilitre units. Medicine bottles are in six sizes from 50 to 500 millilitres. The metric equivalents of the old apothecaries' units are given below:

Capacity

1 minim		=	0·059 ml
1 fluid scruple	= 20 minims	=	1·184 ml
1 fluid drachm	= 60 minims	=	3·552 ml
1 fluid ounce	= 8 drachms	=	28·412 ml

Weight

1 grain		=	0·0648 g
1 scruple (℈)	= 20 grains	=	1·296 g
1 drachm (ʒ)	= 3 scruples	=	3·888 g
1 ounce (℥) (apoth.)	= 8 drachms	=	31·1035 g
1 pound (apoth.)	= 12 ounces	=	373·24 g

Troy Weight

The only unit of troy weight which is now legal for use in trade in the United Kingdom is the troy ounce. Precious metals are weighed in multiples and decimals of this unit.

(The troy ounce is equal to the apothecaries' ounce.)

1 Grain		=	0·0648 g
1 Pennyweight (dwt)	= 24 grains	=	1·552 g
1 troy ounce	= 20 dwt	=	31·1035 g

Nautical Measure

1 fathom	=	6·08 feet
1 cable's length	=	100 fathoms
1 nautical mile	=	10 cables' lengths 1000 fathoms 6080 feet

continued

Nautical Measure continued

1 degree (of latitude) = 60 nautical miles (about $69\frac{1}{2}$ miles)

Circumference of the Earth = 360 degrees

1 international nautical mile = 1852 m

(1 U.K. nautical mile = 1853·18 m)

Hay and Straw Weight

1 truss = { 36 pounds of straw; 56 pounds of old hay; 60 pounds of new hay }

1 load = { 36 trusses; 11 cwt 64 lb of straw; 18 cwt of old hay; 19 cwt 32 lb of new hay }

Hay is reckoned as 'old' after 29th September.

Metric Bed Sizes

Standard Single

200 cm × 100 cm (6 ft $6\frac{3}{4}$ in × 3 ft $3\frac{3}{8}$ in)

Small Single

190 cm × 90 cm (6 ft $2\frac{3}{4}$ in × 2 ft $11\frac{1}{2}$ in)

Standard Double

200 cm × 150 cm (6 ft $6\frac{3}{4}$ in × 4 ft 11 in)

Small Double

190 cm × 135 cm (6 ft $2\frac{3}{4}$ in × 4 ft 5 in)

TIME AND CALENDAR

MEASUREMENT OF TIME

1 Minute	= 60 Seconds
1 Hour	= 60 Minutes
1 Day (Mean Solar Day)	= 24 Hours
1 Week	= 7 Days
1 Lunar Month	= 28 Days
1 Calendar Month	= 28, 29, 30, or 31 Days
1 Year	= 12 Calendar Months 52 weeks 365 Days
1 Leap Year	= 366 Days
1 Century	= 100 Years

A Sidereal Day is 23 hours, 56 min, 4·09 sec, being the time that elapses between two successive passages of a ***fixed star*** over the meridian.

A Solar Day is 24 hours, 3 min, 56·55 sec of sidereal time, and during this interval the earth rotates through 360°59′ 8·33″.

A Solar or **Tropical Year** is 365 days, 5 hours, 48 min, 46 sec, being the length of time in which the earth performs its orbital revolution round the sun.

A Sidereal Year is 365 days, 6 hours, 9 min, 9 sec, being the interval between the sun's apparently leaving a fixed point or star and returning to it.

A Lunar Month in astronomy is 27 days, 7 hours, 43 min, 11·461 sec, being the time during which the moon revolves round the earth.

A Solar Month is the period in which the sun passes through a sign of the zodiac, or the twelfth part of the heavens.

The Hours of the Day

For everyday use the twenty-four hours of a day are divided into two periods of twelve hours, referred to as a.m. (before noon) and p.m. (after noon). The *'twenty-four hour clock'* is also used, in particular for some timetables and for astronomical and scientific purposes, where the hours are numbered from 0 to 23, beginning at midnight.

The Calendar Months

January (31 days); February (28); March (31); April (30); May (31); June (30); July (31); August (31); September (30); October (31); November (30); December (31). Every fourth year is a Leap Year and has 366 days, the extra one being February 29th. But the last year of a century is not a Leap Year unless its number is divisible by 400, e.g. 1900 was not, but 2000 will be.

The Seasons

Spring commences . .	March 21	Autumn commences . .	Sept. 23
Summer commences . .	June 22	Winter commences . .	Dec. 22

The Quarter Days

England, Wales, and Northern Ireland

Lady Day . . . . March 25
Midsummer . . . . June 24
Michaelmas . . . . September 29
Christmas . . . . December 25

Scotland (Term Days)

Candlemas . . . . February 2
Whitsun . . . . May 15
Lammas . . . . August 1
Martinmas . . . . November 11

STANDARD

The time indicated by the sun changes by 1 hour for every 15° of longitude, which is approximately 5 seconds to the mile in middle latitudes.

Greenwich Mean Time (Universal Time) is the time shown by the sun as observed at Greenwich, London.

Almost all countries have adopted a Standard Time which differs from that of Greenwich by an integral number of

COMPARISON OF STANDARD
(as at 12 noon

Accra Ghana 12 noon 12.00 hours	**Adelaide** S. Australia 9.30 p.m. 21.30 hours	**Algiers** Algeria 12 noon 12.00 hours	**Athens** Greece 2 p.m. 14.00 hours
Cairo Egypt 2 p.m. 14.00 hours	**Calcutta** India 5.30 p.m. 17.30 hours	**Canberra** N.S.W. Australia 10 p.m. 22.00 hours	**Cape Town** South Africa 2 p.m. 14.00 hours
Geneva Switzerland 1 p.m. 13.00 hours	**Gibraltar** 1 p.m. 13.00 hours	**Havana** Cuba 8 a.m. 08.00 hours	**Hong Kong** 8 p.m. 20.00 hours
Madrid Spain 1 p.m. 13.00 hours	**Melbourne** Victoria, Australia 10 p.m. 22.00 hours	**Montreal** Quebec, Canada 7 a.m. 07.00 hours	**Moscow** U.S.S.R. 3 p.m. 15.00 hours
Paris France 1 p.m. 13.00 hours	**Peking** China 8 p.m. 20.00 hours	**Perth** W. Australia 8 p.m. 20.00 hours	**Prague** Czechoslovakia 1 p.m. 13.00 hours
San Francisco California, U.S.A. 4 a.m. 04.00 hours	**Santiago** Chile 8 a.m. 08.00 hours	**Singapore** 7.30 p.m. 19.30 hours	**Stockholm** Sweden 1 p.m. 13.00 hours

TIME

hours. Large countries such as the United States of America, Canada, Australia and the U.S.S.R. are divided into time zones. The United Kingdom observes **G.M.T.** in the winter months. In the summer months, between dates which are decided annually, the United Kingdom observes **British Summer Time**, which is 1 hour fast on G.M.T. and the same as **Central European Time**.

TIMES AROUND THE WORLD
Greenwich Mean Time)

Auckland New Zealand 12 midnight 24.00 hours	**Bonn** West Germany 1 p.m. 13.00 hours	**Brussels** Belgium 1 p.m. 13.00 hours	**Buenos Aires** Argentine 8 a.m. 08.00 hours
Colombo Ceylon 5.30 p.m. 17.30 hours	**Dacca** East Pakistan 6 p.m. 18.00 hours	**Delhi** India 5.30 p.m. 17.30 hours	**Dublin** Eire 12 noon 12.00 hours
Karachi West Pakistan 5 p.m. 17.00 hours	**Kingston** Jamaica 7 a.m. 07.00 hours	**Lima** Peru 7 a.m. 07.00 hours	**London** United Kingdom 12 noon 12.00 hours
Nairobi Kenya 3 p.m. 15.00 hours	**New York** U.S.A. 7 a.m. 07.00 hours	**Nicosia** Cyprus 2 p.m. 14.00 hours	**Omsk** U.S.S.R. 6 p.m. 18.00 hours
Reykjavik Iceland 12 noon 12.00 hours	**Rio de Janeiro** Brazil 9 a.m. 09.00 hours	**St. John** N.B., Canada 8 a.m. 08.00 hours	**Salisbury** Rhodesia 2 p.m. 14.00 hours
Tokyo Japan 9 p.m. 21.00 hours	**Vancouver** B.C., Canada 4 a.m. 04.00 hours	**Vladivostok** U.S.S.R. 10 p.m. 22.00 hours	**Washington** D.C., U.S.A. 7 a.m. 07.00 hours

GENERAL INFORMATION

Common Abbreviations

a.c. alternating current
a/c account
A.D. (*anno Domini*), in the year of our Lord
ad lib. (*ad libitum*), at pleasure
A.F.C. Air Force Cross
a.m. (*ante meridiem*), before noon
anon. anonymous
ASDIC Allied Submarine Detection Investigation Committee
A.W.O.L. absent without leave
B.A. Bachelor of Arts
Bart. (or Bt.) Baronet
B.B.C. British Broadcasting Corporation
B.C. before Christ
b.e. bill of exchange
B/L bill of lading
B.M.A. British Medical Association
B.Sc. Bachelor of Science
B.S.T. British Summer Time
C. Centigrade; Celsius; Conservative
c. (*circa*), about
© copyright
C.B.E. Commander (of the Order) of the British Empire
cf. *confer* (compare)
C.G.S. centimetre-gramme-second
C.H. Companion of Honour
C.I.D. Criminal Investigation Department
c.i.f. cost, insurance, freight
c/o care of
C.O.D. cash on delivery
C.S.E. Certificate of Secondary Education
cwt hundredweight
d.c. direct current
del. delegate; (*delineavit*) (he, or she, drew it)
D.F.C. Distinguished Flying Cross
do. ditto; the same
D.S.O. Distinguished Service Order
E. & O.E. errors and omissions excepted
E.E.C. European Economic Community
EFTA European Free Trade Association
e.g. (*exempli gratia*), for example
E.R. Elizabeth Regina; Queen Elizabeth
ERNIE electronic random number indicating equipment
et al. (*et alii*), and others
etc. (et cetera), and the others; and so forth
et seq. (*et sequen/s*), and the following
F. Fahrenheit
f.a.s. free alongside ship
F.D. *fidei defensor* (Defender of the Faith)
f.o.b. free on board
GATT General Agreement on Tariffs and Trade
G.C. George Cross
G.C.E. General Certificate of Education
G.M. George Medal
G.M.T. Greenwich Mean Time
h.c.f. highest common factor
H.M. Her (His) Majesty
H.M.S. Her (or His) Majesty's Service or Ship
h.p. horse-power
ib. (or ibid.) (*ibidem*), in the same place
I.B.A. Independent Broadcasting Authority
I.C.B.M. intercontinental ballistic missile
id. (*idem*), the same
i.e. (*id est*), that is

I.M.F. International Monetary Fund
incog. (*incognito*), with concealed identity; in secret
in loco in place of
I.Q. intelligence quotient
ISBN International Standard Book Number
i.t.a. initial teaching alphabet
J.P. Justice of the Peace
K.B.E. Knight (Commander) of the British Empire
K.C.B. Knight Commander of the Bath
K.G. Knight of the (Order of the) Garter
L. Liberal
Lab. Labour (party)
lat. latitude
lb (*libra*), pound weight
l.c.m. least common multiple
long. longitude
M.A. Master of Arts
M.B.E. Member of (the Order of) the British Empire
M.C. Military Cross; Master of Ceremonies
M.M. Military Medal
M.P. Member of Parliament; Military Police
m.p.h. miles per hour
MS. (*pl.* MSS.) manuscript
M.Sc. Master of Science
NATO North Atlantic Treaty Organization
N.B. (*nota bene*), mark well
n.d. no date (of books)
net not subject to deduction
No. (*numero*), number
ob. (*obiit*), died
O.B.E. Officer (of the Order) of the British Empire
O.M. (Member of the) Order of Merit
oz ounce
p.c. per cent
P.C. police constable; Privy Council, -lor
Ph.D. Doctor of Philosophy
p.m. (*post meridiem*), afternoon
P.O. The Post Office
p.p. (or *per pro.*) (*per procurationem*), by proxy
pro tem. (*pro tempore*), for the time being
PS. (*postscriptum*), postscript
P.T.O. please turn over
Q.C. Queen's Counsel
Q.E.D. (*quod erat demonstrandum*), which was to be proved
q.v. (*quod vide*), which see
R.C. Roman Catholic
R/D Refer to drawer (banking)
R.I.P. (*Requiescat in pace*), May he (she, or they) rest in peace!
r.m.s. root mean square
r.p.m. revolutions per minute
R.S.V.P. (*répondez, s'il vous plaît*), Answer if you please
SI (*Système International d'Unités*), International System of (Metric) Units
sic so written
SOS ('Save our Souls'), distress signal
sp. gr. specific gravity
STD Subscriber Trunk Dialling
stet let it stand
UHF ultra high frequency
U.N. United Nations
UNESCO United Nations Educational, Scientific, and Cultural Organization
v. (versus), against
V.C. Victoria Cross
VHF very high frequency
viz. (*videlicet*), namely
W.H.O. World Health Organization

The Compass

This instrument is used to indicate the line of the magnetic meridian. In the **Mariner's Compass** there is a magnetised needle which points to the North and a card printed with the points of the compass. When the needle and the North on the card are in line the bearing of any object or the direction of a course can be ascertained. The magnetic needle is affected by electrical or magnetic disturbances, and **gyrostatic compasses** are now widely used in aircraft and ships. These use the property that the axis of rotation of a rapidly spinning object tends to stay pointing in the same direction.

The 32 points of the compass are shown on the diagram. The angle between two adjacent points is 11° 15′. In modern navigation, bearings are quoted in degrees east of north, e.g. 135° instead of SE, 270° instead of W.

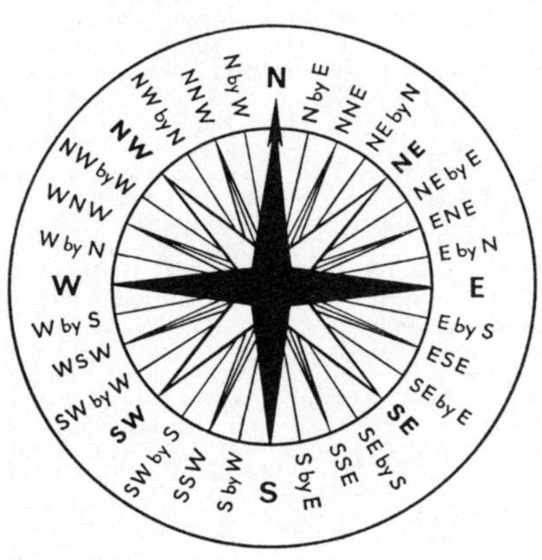

The Barometer

The barometer is an instrument for measuring atmospheric pressure. The standard method employed is to balance a column of mercury against a column of air. The height of the mercury column supporting the air column is taken as the pressure at the time. Pressure can be quoted in millimetres or inches of mercury but it is commonly expressed in millibars. The SI unit N/m^2 is likely to supersede these units.

1 Standard Atmosphere = 101 325 N/m^2
= 1013·25 mb = 760 mmHg = 29·92 inHg

With the barometer's aid, changes in the weather can be forecast to some extent. Falling pressure indicates changeable weather and rising pressure indicates fine weather. Atmospheric pressure decreases with increasing altitude, and therefore the barometer can be used as a rough guide to height above sea level. The **aneroid** barometer works by measuring the movements of the lid of a metal box, which is partially exhausted of air, when the atmospheric pressure changes. It is the basic component of an altimeter.

The Thermometer

The thermometer is used to measure the temperature of any object. The most familiar kind consists of a glass tube with a very small diameter containing, in general, mercury or alcohol. The liquid expands or contracts with variations in temperature, and the length of the thread of mercury or alcohol gives the temperature reading on a scale graduated in degrees. Modern forms of the thermometer use thermoelectric devices.

The Elements and their Chemical Symbols

Actinium . . . . Ac
Aluminium . . . Al
Americium . . . Am
Antimony . . . Sb
Argon . . . . . . Ar
Arsenic . . . . . As
Astatine . . . . . At

Barium . . . . . Ba
Berkelium . . . Bk
Beryllium . . . . Be
Bismuth . . . . . Bi
Boron . . . . . . B
Bromine . . . . Br

Cadmium . . . . Cd
Caesium . . . . Cs
Calcium . . . . . Ca
Californium . . Cf
Carbon C
Cerium . . . . . Ce
Chlorine Cl
Chromium . . Cr
Cobalt . . . . . Co
Copper . . . . . Cu
Curium . . . . . Cm

Dysprosium . . Dy

Einsteinium . . Es
Emanation . . . Em
Erbium . . . . . . Er
Europium . . . . Eu

Fermium . . . . Fm
Fluorine . . . . . F
Francium . . . . Fr

Gadolinium . . Gd
Gallium . . . . . Ga
Germanium . . Ge
Gold . . . . Au

Hafnium . . . . Ha
Helium . . . . . He
Holmium . . . Ho
Hydrogen . . . H

Indium . . . . . . In
Iodine I
Iridium . . . . . . Ir
Iron . . . . Fe

Krypton . . . . Kr

Lanthanum . . La
Lawrencium . Lr
Lead . . . . Pb
Lithium Li
Lutetium . . . . Lu

Magnesium . Mg
Manganese . . Mn
Mendelevium . Md
Mercury . . . . . Hg
Molybdenum . Mo

Neodymium . . Nd
Neon . . . . Ne
Neptunium . . Np
Nickel . . . . . . Ni
Niobium . . . . Nb
Nitrogen . . . . N
Nobelium . . . No

Osmium . . . . . Os
Oxygen . . . . . O

Palladium . . . Pd
Phosphorus . . P
Platinum . . . . Pt
Plutonium . . . Pu
Polonium . . . . Po
Potassium . . . K

Praseodymium Pr
Promethium . . Pm
Protactinium . Pa

Radium . . . . . Ra
Rhenium . . . . Re
Rhodium . . . . Rh
Rubidium . . . . Rb
Ruthenium . . . Ru

Samarium . . . Sm
Scandium . . . Sc
Selenium . . . . Se
Silicon . . . . . . Si
Silver . . . . Ag
Sodium . . . . . Na
Strontium . . . Sr
Sulphur . . . . . S

Tantalum . . . . Ta
Technetium . . Tc
Tellurium . . . . Te
Terbium . . . . . Tb
Thallium . . . . Tl
Thorium . . . . . Th
Thulium . . . . . Tm
Tin . . . . Sn
Titanium . . . . Ti
Tungsten . . . W

Uranium . . . . U

Vanadium . . . V

Xenon . . . . . . Xe

Ytterbium . . . Yb
Yttrium . . . . . Y

Zinc . . . . Zn
Zirconium . . . Zr

Greek Alphabet

Α	α	alpha	Ι	ι	iota	Ρ	ρ	rho
Β	β	beta	Κ	κ	kappa	Σ	σ	sigma
Γ	γ	gamma	Λ	λ	lambda	Τ	τ	tau
Δ	δ	delta	Μ	μ	mu	Υ	υ	upsilon
Ε	ε	epsilon	Ν	ν	nu	Φ	φ	phi
Ζ	ζ	zeta	Ξ	ξ	xi	Χ	χ	chi
Η	η	eta	Ο	ο	omicron	Ψ	ψ	psi
Θ	θ	theta	Π	π	pi	Ω	ω	omega

Russian Alphabet

Printed	*Approximate pronunciation*	*Printed*	*Approximate pronunciation*
А а	*a* in 'father'	Р р	*r* in 'rat'
Б б	*b* in 'box'	С с	*s* in 'sat'
В в	*v* in 'vat'	Т т	*t* in 'top'
Г г	*g* in 'give'	У у	*oo* in 'boot'
Д д	*d* in 'day'	Ф ф	*f* in 'fire'
Е е	*ye* in 'yet'	Х х	*kh* as in Scottish 'loch'
Ё ё	*yo* in 'yonder'		
Ж ж	*zh* in 'pleasure'	Ц ц	*tz* in 'cats'
З з	*z* in 'zebra'	Ч ч	*ch* in 'chain'
	s in 'please'	Ш ш	*sh* in 'short'
И и	*ee* in 'meet'	Щ щ	*shch* in 'fresh cheese'
Й й	*y* in 'boy'		
К к	*k* in 'king'	Ъ ъ	indicates hard sign
Л л	*l* in 'love'	Ы ы	*i* in 'fit'
М м	*m* in 'man'	Ь ь	indicates soft sign
Н н	*n* in 'nut'	Э э	*e* in 'men'
О о	*o* in 'hot'	Ю ю	*u* in 'unite'
П п	*p* in 'pen'	Я я	*ya* in 'yard'

The Beaufort Scale of Wind Force

Beaufort number	*Wind*	*Speed* *m.p.h.*	*knots*	*km/h*	*Effect on land*
0	Calm	less than 1	less than 1	less than 2	Smoke rises vertically
1	Light air	1–3	1–3	2–5	Direction shown by smoke but not by wind vanes
2	Light breeze	4–7	4–6	6–12	Wind felt on face; leaves rustle; wind vanes move
3	Gentle breeze	8–12	7–10	13–20	Leaves and twigs in motion; wind extends light flag
4	Moderate breeze	13–18	11–16	21–29	Raises dust, loose paper and moves small branches
5	Fresh breeze	19–24	17–21	30–39	Small trees in leaf begin to sway
6	Strong breeze	25–31	22–27	40–50	Large branches in motion; whistling in telegraph wires; difficulty with umbrellas
7	Moderate gale	32–38	28–33	51–62	Whole trees in motion; difficult to walk against wind
8	Fresh gale	39–46	34–40	63–74	Twigs break off trees; progress impeded
9	Strong gale	47–54	41–47	75–87	Slight structural damage occurs; chimney pots and slates blown off
10	Whole gale	55–63	48–56	88–102	Trees uprooted and considerable structural damage
11	Storm	64–75	57–65	103–120	Widespread damage, seldom experienced in the United Kingdom
12	Hurricane	above 75	above 65	above 120	Winds of this force only encountered in tropical revolving storms

General Index of Retail Prices

1963–1972
Monthly Average for All Items

1963	103·6	1968	125·0
1964	107·0	1969	131·8
1965	112·1	1970	140·2
1966	116·5	1971	153·4
1967	119·4	1972	164.3

(Figures from the Central Statistical Office, Monthly Digest of Statistics)

The World

Area of the Earth	196 836 000 sq. miles	510 100 000 km²
Area of Land	55 786 000 sq. miles	149 000 000 km²
Area of Water	141 050 000 sq. miles	361 000 000 km²
Volume of the Earth	260 000 000 cu. miles	1 083 000 000 km³
Equatorial circumference	24 901·8 miles	40 076 km
Equatorial radius	3 963·3 miles	6 378·2 km

The land of the world is divided as follows:

	sq. miles	*km²*
Europe	1 903 000	4 929 000
Asia (excluding U.S.S.R.)	10 661 000	27 611 000
U.S.S.R.	8 649 000	22 402 000
Africa	11 683 000	30 258 000
North America	8 000 000	20 720 000
South America	6 800 000	17 612 000
Polar Regions	5 000 000	12 950 000
Oceania	3 286 000	8 510 000

The Oceans

Ocean	*Area*		*Greatest Depth*	
	sq. miles	*km²*	*ft*	*m*
Pacific	63 986 000	165 722 000	36 198	11 033
Atlantic	31 530 000	81 662 000	27 498	8 381
Indian	28 350 000	73 426 000	26 400	8 047
Arctic	5 541 600	14 353 000	17 850	5 441
Antarctic	2 000 000	5 180 000	14 271	4 350

Largest Islands

	sq. miles	*km²*
Australia	2 974 581	7 704 100
Greenland	826 000	2 139 300
New Guinea	311 000	805 500
Borneo	284 000	735 600
Madagascar	241 094	624 430
Baffin Land	231 000	598 300
Sumatra	163 000	422 200
Honshu (Japan)	88 911	230 280
Great Britain	88 747	229 850
Celebes	73 000	189 070
Prince Albert	60 000	155 400
South Island (N.Z.)	58 092	150 460
Java	48 400	125 360
North Island (N.Z.)	44 281	114 690
Cuba	44 164	114 380
Newfoundland	42 734	110 680
Ellesmere (Canada)	41 000	106 200

Largest Seas

	sq. miles	*km²*
Malay	3 137 000	8 125 000
Central American	1 770 170	4 584 700
Mediterranean	1 145 000	2 965 500
Bering	878 000	2 274 000
Okhotsk	582 000	1 507 000
East China	480 000	1 243 000
Hudson Bay	472 000	1 222 000
Japan	405 000	1 049 000
Andaman	305 000	790 000
North Sea	221 000	572 000
Red Sea	178 000	461 000
Baltic	158 000	409 000

Longest Rivers

River	*Flowing into*	*miles*	*km*
Mississippi–Missouri	Gulf of Mexico	4500	7200
Amazon	Atlantic	4000	6400
Nile	Mediterranean	4000	6400
Yangtse	North Pacific	3400	5500
Yenisei	Arctic Ocean	3300	5300
Congo	Atlantic	3000	4800
Lena	Arctic Ocean	3000	4800
Mekong	South China Sea	2800	4500
Obi	Arctic Ocean	2700	4300
Niger	Gulf of Guinea	2600	4200
Hoangho	North Pacific	2600	4200
Amur	North Pacific	2500	4000
Parana	Atlantic	2450	3900
Volga	Caspian Sea	2400	3900
Mackenzie	Beaufort Sea	2300	3700

The Severn is the longest river in Great Britain, flowing 220 miles, 350 km, into the Bristol Channel. The longest river in England, the Thames, flows 210 miles, 340 km, to the North Sea.

Highest Mountains

Name	*Range (or Country)*	*ft*	*m*
Everest	Himalayas	29 002	8840
Godwin–Austen, K2	Karakoram	28 250	8611
Kinchinjunga	Himalayas	28 146	8579
Makalu	Himalayas	27 790	8470
Dhaulagiri	Himalayas	26 800	8169
Nanga–Parbat	Himalayas	26 660	8126
Nanda Devi	Himalayas	25 650	7818

continued overleaf

Name	*Range (or Country)*	*ft*	*m*
Kamet	Himalayas	25 450	7757
Tirich Mir	Afghanistan	25 400	7742
Ulugh Muztagh	Tibet	25 300	7711
Illampu	Andes	25 248	7696
Illimani	Andes	24 633	7508
Jonsong	Himalayas	24 344	7420
Abi Gamin	Himalayas	24 170	7367

Ben Nevis (Scotland), 4406 feet, 1343 m, is the highest mountain in Great Britain. Snowdon (Wales) is 3560 feet, 1085 m, and Scafell Pike (England), 3210 feet, 978 m.

Largest Lakes

		sq. miles	*km²*
Caspian Sea	Asia	170 000	440 300
Superior	N. America	31 800	82 400
Victoria	Africa	26 200	67 900
Sea of Aral	Asia	24 400	63 200
Huron	N. America	23 010	59 600
Michigan	N. America	22 400	58 000
Chad	Africa	20 000	51 800
Nyasa	Africa	14 200	36 800
Tanganyika	Africa	12 700	32 900
Baikal	Siberia	11 580	30 000
Great Bear	Canada	11 200	29 000
Great Slave	Canada	11 170	28 900
Erie	N. America	9 960	25 800
Winnipeg	Canada	9 459	24 500
Ontario	N. America	7 540	19 500

Loch Lomond (Scotland), about 24 miles, 39 km, in length and of varying breadth, is the largest in Great Britain. In the Lake District of England, Windermere, about 10 miles, 16 km, long is the largest, with Ullswater and Derwentwater next.

The Planets

		Distance from the sun	
		millions of miles	*millions of km*
Mercury	☿	36	58
Venus	♀	67	108
Earth	⊕	93	150
Mars	♂	142	228
Jupiter	♃	484	778
Saturn	♄	887	1427
Uranus	♅	1795	2870
Neptune	♆	2797	4497
Pluto	♇	3670	5900

The Moon

Diameter 2161 miles, 3476 km
Distance from Earth 238 840 miles, 384 400 km
Phases of the Moon ● New moon ○ Full moon
☽ First quarter ☾ Last quarter

Velocity of light 186 325 miles per second
299 792·5 km per second

Velocity of sound 1087 ft per second, 331 m per second
762 miles per hour, 1193 km per hour

Signs of the Zodiac

♈	Aries, the Ram	♎	Libra, the Balance
♉	Taurus, the Bull	♏	Scorpio, the Scorpion
♊	Gemini, the Twins	♐	Sagittarius, the Archer
♋	Cancer, the Crab	♑	Capricornus, the Goat
♌	Leo, the Lion	♒	Aquarius, the Waterbearer
♍	Virgo, the Virgin	♓	Pisces, the Fishes

INTERNATIONAL ORGANISATIONS

United Nations

Principal Organs General Assembly; Security Council; Economic and Social Council; Trusteeship Council; International Court of Justice; Secretariat.

Principal Commissions and Bodies within the Economic and Social Council

Regional Economic Commissions: Europe (**ECE**); Asia and the Far East (**ECAFE**); Latin America (**ECLA**); Africa (**ECA**).

Functional Commissions: Disarmament; Statistics; Population; Social; Human Rights; Status of Women; Narcotic Drugs.

Special Bodies UN Children's Fund (**UNICEF**) Commissioner for Refugees Conference on Trade and Development (**UNCTAD**) Industrial Development Organisation (**UNIDO**)

Intergovernmental Agencies (previously called Specialised Agencies) International Atomic Energy Agency (**IAEA**) International Labour Organisation (**ILO**) Food and Agriculture Organisation (**FAO**) UN Educational, Scientific and Cultural Organisation (**UNESCO**) World Health Organisation (**WHO**) World Bank (**Bank**) International Finance Corporation (**IFC**) International Monetary Fund (**Fund**) International Development Association (**IDA**) International Civil Aviation Organisation (**ICAO**) Universal Postal Union (**UPU**) International Telecommunication Union (**ITU**) World Meteorological Organisation (**WMO**) Intergovernmental Maritime Consultative Committee (**IMCO**) General Agreement on Tariffs and Trade (**GATT**)

Western Europe

European Economic Community (EEC) (Common Market) (Treaty of Rome, 1957)
European Coal and Steel Community (ECSC) (1952)
European Atomic Energy Community (EURATOM) (1957)

Members Belgium, Denmark, France, Germany, The Republic of Ireland, Italy, Luxembourg, The Netherlands, The United Kingdom. The three communities now share a Commission, Council of Ministers, Parliament and Court of Justice.

European Free Trade Association (EFTA) (1959)

Members Austria, Finland, Iceland, Norway, Portugal, Sweden, Switzerland.

North Atlantic Treaty Organisation (NATO) (1949)

Members Belgium, Canada, Denmark, Greece, Iceland, Italy, Luxembourg, the Netherlands, Norway, Portugal, Turkey, the United Kingdom, the United States of America and the Federal Republic of Germany. (France withdrew in 1966.)

Other Groupings of States

Warsaw Pact (Eastern European Mutual Assistance Treaty) (1955)

Comecon (Council for Mutual Economic Assistance) (1949)

Eastern Joint Institute for Nuclear Research (1956)

Members The Soviet Union and certain other Communist countries.

Afro-Asian People's Solidarity Council (1957)
Afro-Asian Organisation for Economic Co-operation (1958)
Three Continents Solidarity Organisation (1966)
Members 'Non-aligned' or 'Uncommitted' Nations.

Organisation of American States (OAS) (1948)
Organisation of Central American States (OCAS) (1951)
Latin American Free Trade Association (LAFTA) (1960)

The Arab League

Central Treaty Organisation (CENTO)
Members The United Kingdom, Iran, Pakistan and Turkey (Ass. member USA).

Organisation of African Unity (OAU) (1963)
Members All independent countries in Africa except South Africa.

The Anzus Pact (Pacific Security Treaty) (1952)
Members Australia, New Zealand and the United States of America.

South-East Asia Treaty Organisation (SEATO)
Members Australia, the United Kingdom, France, New Zealand, Pakistan, the Philippines, Thailand and the United States of America.

Colombo Plan—a plan for the economic development of South and South-East Asia (1951)